ALIEN ABDUCTION
OF
THE WYOMING HUNTER

First person story of Carl Higdon
October 25, 1974

MARGERY HIGDON

The author has recreated events, locales, and conversations from her memories of them. In order to maintain their anonymity in some instances, the author may have changed the names of individuals and places to protect the privacy of the individuals.

All rights reserved. No part of this book may be reproduced in any form or by any means without the written permission of the author, except brief quotes used in reviews.

First printing: December 2017
Volume One

Copyright © 2017 by Margery Higdon
Cover design by Margery Higdon & R. L. Copple

ISBN Number: 13: 978-1981812899
ISBN Number: 10: 198181289X

ALIEN ABDUCTION
OF
THE WYOMING HUNTER

First person story of Carl Higdon

October 25, 1974

CHAPTER ONE

It is hard to believe that this year, 2017, has been 43 years since that, **out of this world,** Elk hunting trip...

October 25, 1974

Carl went to pick up his crew to go to work. But, this morning his crew members were sick with the flu and couldn't go to work. So... since this was Elk Hunting Season, Carl decided to go hunting.

Carl went home, picked up his *New 7MM Mag rifle,* and headed south of town toward McCarty Canyon.

Carl was on the fork in the road leading into McCarty Canyon when he encountered some fellow hunters. They were working on their pickup. It had stalled, and they couldn't get it to start.

Carl stopped to see if he could assist them. While they worked on the pickup, they discussed hunting and shot the breeze, as hunters do. Carl told them of his plans to go into McCarty Canyon to see what he could see there. But, the hunters suggested that the hunting was much better in the forest.

Carl and the hunters each drank a cup of coffee. The pickup was fixed, and the hunters drove off. Carl decided to head for the forest... instead of going to McCarty Canyon.

WAS THIS DAY PLANNED OR WAS IT CIRCUMSTANCIAL?

CARL HAD PLANNED TO GO HUNTING IN THE CANYON, AND THE HUNTERS SUGGESTED THAT HE GO HUNTING IN THE FOREST...

It was a beautiful autumn day as Carl drove into the forest. He parked his pickup upon a knoll and poured a cup of coffee. He got out and stood in front of his pickup to survey the area; when the Game Warden drove up. Carl and the Game Warden discussed hunting in the area and shot the breeze over coffee.

Carl told the Game Warden he thought he would go down the hill and see if he could find anything...

NEVER SUSPECTING WHAT HE WAS ABOUT TO FIND...

The Game Warden got into his pickup and drove off. Carl put his cup and thermos back into his pickup. He took his *New 7 MM Mag rifle* out of the case. With the rifle under his arm, he left the pickup on the hill and walked down the winding road – winding around the trees, going up and down the hills, following the road.

He walked to the left, going deeper into the trees. He came to a clearing. He stopped and surveyed the area. Five elk stood in the clearing. He raised his rifle to get one in his sights; he got a bull and pulled the trigger...

But, the shot makes no noise! He watches the 7 MM Mag bullet as it comes out the rifle barrel, goes out a few feet — stops in midair, and falls to the ground!

ALL IN SLOW MOTION!

HE WATCHED THE BULLET COME OUT OF THE RIFLE...

How can this be? A 7MM travels over 3600 feet a second... and he WATCHED it come out of the rifle... go a few feet... and fall to the ground! The elk are still

there! There was no noise to scare them! They stand motionless!

There is no noise. The birds are even silent. Everything is totally silent.

WHAT IS GOING ON...?

Carl walked over to the spot where the bullet fell. He bent down, picked it up, and put it in his canteen pouch.

He sensed something behind him. He turned around. There is someone there! It is a man. But, he looked different from any ***man*** Carl had ever seen.

The ***man*** was about six feet tall. He had straw-colored hair – that stood straight up. He had two strands of hair, in the front – that looked like... *antennas*. His face seemed to go back into his neck like he had no chin. His face was yellowish. He was dressed in a black suit, kind of what a scuba diver would wear. His legs were bowed.

The ***man*** asked Carl, "Are you hungry?"

Carl replied, "A little." A packet of pills ***floats*** over to Carl.

The *man* tells Carl to take one. Carl does as *he* says. But, he doesn't know why. He seldom ever takes even aspirin. He doesn't like to take pills. He doesn't know why he took the pill. It was as if he had no willpower.

The *man* then asks Carl, if he wants to go with him.

Carl says, "I might as well…"

CHAPTER TWO

The next thing I know I am inside...

A GLASS CUBICLE.

I will call it that, for I know of no other words to describe it. I am sitting in a high back seat, like a bucket seat. My arms are strapped down. Something is on my head... I can't move!

I can't turn around, but, I sense the five elk behind me. They are reflected in the glass overhead like a mirror. They're in a cage. They are motionless.

The next thing I see is my pickup. I am looking... **down at my pickup sitting on the hill. This *being* waves his hand... and my pickup disappears.**

Next, I look **down** and see a **blue ball. It looks like... a huge marble.**

THEN... Everything... BLACK...

THEN... LIGHT... BRIGHT LIGHT!

🙞 🙞 🙞

I can barely keep my eyes open. The **LIGHT IS SO BRIGHT!** My eyes start to tear profusely. There is a tower where the light is coming from. It is... kind of like the Space Needle at the world's fair. But... the **LIGHT IS BRIGHT!** I can barely see! There, below the tower, are people. People like me! They seem to be at ease and talking amongst themselves.

They – there are now two – take me down a long walkway. We don't walk — we *float!* They take me inside a large room. A wall comes in front of me.

This seems to be... some kind of an examination?

They say, '*I am not what they want. They will take me back.*'

We are then back outside. Again, we do not walk – we **FLOAT**. We are now, again, back in the **LIGHTS!** *OH, THE LIGHTS! THE LIGHTS!* **THEY ARE SO BRIGHT! The light burns my eyes! It's unlike any feeling I've ever had! My eyes keep watering. It's hard for me to see!**

We are now inside the **craft**. This time there is only one **being**. He says his name is **AUZZO ONE**. He will take me back. *I am not what they want.*

Auzzo One tells me, '*They have been coming to earth for many years. They come in search of fish and animals. Their food is in the form of pills. One pill will last for three days. They make the pills from fish and animals. Their ocean is yellow. All the fish have died.*'

He shows me a land mass map of his planet. He tells me, "We are **163,000-light-miles from earth**. The nine planets of our solar system supply the magnetic force for our power."

He takes my rifle. He studies it. He said he would like to keep it. It is a primitive weapon. But, he is not allowed to. **'Primitive... I just bought it!** He tells me that they wear black because our sun burns them. The patches are symbols of their planet. He said he is a hunter.

⚜ ⚜ ⚜

Chapter Three

We are now **over the forest.** Auzzo tells me he will see me. I just... **FLOAT DOWN OUT OF THE CRAFT!** I come down on the side of a hill. My foot slips on a rock, and I fall. **OH! My shoulder! IT HURTS! IT HURTS!**

I don't know where I am. I see a sign. I can barely make it out – **North Boundary Lincoln Forest**. *But, where is that?* I walk down a road to the bottom of the hill. I don't know where I am. It is getting cold and dark. I come upon a pickup (not knowing it is mine). I get in and sit. I hear people talking. I finally figure that it is coming from the radio, a two-way radio. After some difficulty, I figure out how to use the radio. I talk to the person on the other end. He wants to know... who I am and where I am.

I don't know! *Who am I? Where am I?*

I tell this person that I saw a sign that read:

NORTH BOUNDARY LINCOLN FOREST

'*But, I don't know where that is. I am sitting in a pickup with a funny stick in the middle. I don't know what it's for.* (Carl learned how to drive with a stick shift when he was young.)

The person on the radio tells me, "Look in the glove compartment for some papers, take the papers out; then read out-loud what is on the papers."

He keeps talking to me, and asking, "Who are you? Where are you?"

I keep telling him, "I don't know. I don't know. I am so cold."

WHERE ARE MY ELK? MY ELK,

WHERE ARE MY ELK?

MY PICKUP — My pickup just disappeared! The big blue ball...!

"The men in black...!"

"I'm so cold..."

I keep saying, "I don't know. I don't know. I am so cold... so cold."

ᐃᐃᐃ

Chapter Four

Margery came home around 4:00 p.m. from her job with *City Plumbing & Heating*. She had this uneasy feeling that she needed to find Carl. She knew that Carl had gone hunting out south, but she didn't know where. This feeling that she must go to Carl kept getting stronger and stronger. Finally, she called her girlfriend, Marilyn. She told her of her uneasy feeling and asked if she would take her out to look for Carl.

The feeling is so strong; she feels that she needs to find him – **now**.

Marilyn explained to Margery, that if Carl bagged an elk, it would take him some time to field dress and carry it out... Marilyn told her to just be patient.

Margery knew this to be true; but, she still had this uneasy feeling that she **needed** to go to Carl — **now**.

She decided to busy herself and started supper for the family. As she worked with supper, the feeling eased.

Around 6:30 p.m. the phone rang. Bud Rosaker is on the other end. He wanted to know where Carl is. Margery told

Bud she really doesn't know. Shortly the phone rang again. This time it is the company office in Riverton, Wyoming. It is Carl's head boss, Andy Anderson.

He said, "We know Carl went hunting. We need to know where he went. We have someone on the company radio. We think this man has been hurt. It sounds like Carl. But, we aren't for sure. Now, where did he go hunting?"

Margery told Andy she really didn't know where Carl went. She just knew he had talked about going hunting.

☙ ☙ ☙

Chapter Five

The office in Riverton has established that it is Carl's company pickup, and the man on the radio is **probably** Carl.

The sheriff said that the Game Warden talked with Carl earlier this afternoon. Carl had been standing in front of his pickup, parked on a knoll. So, now they have narrowed the radio call down to the forest south of town, and have pinpointed the general area.

The sheriff had requested an aerial search, but the pilot had drunk an alcoholic drink and couldn't go up. The sheriff has now organized a ground search party.

&&&

The search party consists of experienced hunters with several four-wheel drive vehicles.

As the search party gassed up their vehicles, Bud called Margery to let her know that they are on their way to find Carl. Margery told Bud that she wants to go with them.

Bud says, "If you don't have a four-wheel vehicle there's no way you can get into that area."

Once again, Margery called her friend, Marilyn. "Can you and Don take me out to find Carl? The sheriff has organized a search party, and they will go as soon as they get gas."

Marilyn told her, "Don and I will be right down to get you. We will need to get gas also."

Chapter Six

Don and Marilyn filled their pickup with gas and put in extra cans of gas. They picked Margery up at her house and headed south for the forest. By this time, the search party had already left.

Each of the pickups in the search party had *CB Radios*. Don also had a *CB* in his pickup. As Don, Marilyn, and Margery enter the forest, Don called on the *CB* to see if they could see Carl yet?

A reply came back, "See him? **HELL**, we aren't even into the area yet! We keep getting stuck and have to pull each other to keep going. It's a **HELL** of a mess in here!"

"Stay on top of the hill. Don't even think of coming in here! We will call you when we find him. His office still has him on the company radio. We are – pretty sure – of the area."

Margery, Marilyn, and Don sit in the pickup on the top of the hill. They are anxiously waiting for communication from the search party. Waiting... for them to find Carl. Just waiting... It seemed like it already had been so... long.

The night is BRIGHT! But, there is no full moon. It is so light out; you could almost pick up a dime on the road.

After what seems like hours, Don again called to the search party. He asked if they had come to him yet. The reply came back, **"Hell NO!** We are still getting stuck. The road is narrow and curves around the trees. It is a mess. It is slow going. **We will call you when we get to him!"**

"STAY WHERE YOU ARE! STAY — WHERE YOU ARE!"

☙☙☙

Chapter Seven

Marilyn, Don, and Margery decide they may as well try to get some rest while they waited. Hoping to catch some zzz's, Marilyn laid her head on the back of the seat. Suddenly, Marilyn screamed, "Quit moving the truck!"

Don and Margery, both, at the same time, said, "We aren't moving the truck. What's wrong with you?"

Marilyn exclaimed, "Look at that star! It is moving and changing colors and doing crazy things!"

Don, tired and annoyed, told her, "Close your eyes and go back to sleep. You must be dreaming."

After that, they all laid their heads back and tried to get some rest; when the search party called over the CB, "We have Carl. But, we are going out a different way. It will be a lot easier to go out by the gate at the bottom. Meet us there."

Don started the pickup, turned around and drove to the gate at the bottom of the hill, near the entrance to the forest. He killed the motor and the three set up their vigil of waiting. This time, they are waiting for the search party to bring him out.

CHAPTER EIGHT

Over to the East, you can see the lights of the pickups as they snake their way, along the winding road. Over the hill, farther to the East, is a bright red-orange light. It looks like... a sunrise.

None in the party even gave any consideration to the fact it is late-night, and it is **way too early** for the sun to rise. They don't even give it another thought.

Finally, the search party reached the gate. Someone in the party got out of their pickup and unlocked the gate. The search party drove through the gate. They all stopped to put gas into their pickups from the cans they brought; while someone relocked the gate.

Margery ran over to the pickup where Carl was. She was so happy to see him. But... he seemed different. She tried to talk to him. He just looked at her – as though he was looking through her.

Finally, she asked him if he got his elk. He looked dazed. He looked **up** through the windshield, and in a sing-song type of voice, he said, "They took my elk. They took my elk."

Carl was shivering. He was so cold. Margery took off her coat and... tried to drape it over his shoulders.

He moved back crying, **"Don't touch me! Don't touch me!"**

Margery was scared. She's not sure what's wrong with her husband.

She told Don, "Get the rifle! Get it out of the truck! I don't know what's wrong with him!"

Don removed the rifle from the gun holder behind Carl and disappeared. Margery again tried to talk to Carl. He just kept looking **up through the windshield, towards the sky**. He kept repeating in a sing-song voice, "They took my elk. They took my elk."

By this time, everyone in the search party had filled their pickups with the extra gas they had brought, and were ready to go. Bud decided that with the condition Carl is in, they had better take him to the hospital.

Don took Margery by the shoulders and led her back to his truck.

The search party, with Don following, headed down the road for the *Carbon County Memorial Hospital*, in Rawlins.

ꙮꙮꙮ

Chapter Nine

Shortly, they meet Roy Fleming driving up the main road leading into the forest. He has driven all the way from Riverton to see if he could help Carl.

Roy stopped Bud's pickup to see if Carl was okay. Roy told Bud that maybe Carl would be more comfortable lying down in the back seat of his car.

Roy got out of his car, went around and opened the back door, to let Carl in.

When *ALL HELL* broke loose! Carl bolts and runs from Bud's pickup! The deputy sheriff runs to the side of the ditch. He crouches down with his gun drawn and steadied on his knee – ready to shoot if needed. Carl is not acting like himself... no one knows what he might do!

Carl is running with his arms over his eyes! He is running and screaming, **"The LIGHTS... OH, GOD... THE LIGHTS!"**

Margery finally realizes that the car lights are what Carl is screaming about. She and Bud yell for everyone to turn off their lights. **"TURN OFF YOUR LIGHTS!"**

Everyone turned their lights off. Carl then calmly walked over to the back door of Roy's car and slammed it shut. He then opened the front door, got in, and shut the door.

Once again, the search party is headed for the hospital in Rawlins.

Carl is sitting in the front seat of Roy's car and talks with him. He wants to know why Roy isn't dressed in black. "Doesn't the sun burn you?" he asked.

Carl talked about his **163,000-light-mile trip into outer space**.

ꙮꙮꙮ

Carl talked of a lot of things about this trip; that to this day, Roy does not want to talk about.

ꙮꙮꙮ

Chapter Ten

Carl is taken to the *Carbon County Memorial Hospital* emergency room in Rawlins, Wyoming. His eyes are watering profusely. His shoulder **hurts**. He keeps asking for his three-day pills, the ones, that... **just float to you**.

Dr. Tongco is on duty. He is a new doctor in town, and Carl is one of his first patients. He is intrigued by what this patient is telling him — About the **Beings** dressed in black... *About his 163,000-light-mile journey... Cone hands... that, when waved, everything disappeared. Not walking... just floating. Just going from here to there... no walking.'*

THE LIGHTS — THE LIGHTS

THEY BURN! — THEY BURN!

Dr. Tongco is so intrigued with his new patient that he makes notes of these references in Carl's medical records.

Carl seems to be in an amnesic state. He doesn't know who he is. When his wife asks him his name, he says, "They keep calling me Carl. Who is Carl?"

His shoulder hurts. His eyes are watering profusely. But, he doesn't appear to have any broken bones or bruises.

Dr. Tongco ordered tests. He also ordered a drug test, to see if Carl is on any drugs. *The way this man is talking he must be on something.*

Nurse Peterson tried to make Carl comfortable. She put a doubled-over wet washcloth over his eyes to help soothe them. She takes her hand and goes over his body, to see where he is hurt. Carl says, "It hurts." But, when she goes back over these areas again, Carl says, "It feels better now."

Carl is complaining about the lights hurting his eyes; even with the doubled-over washcloth over them. So, Nurse Peterson turned off the lights.

Carl's wife, Margery, sits in a chair near his emergency room bed, anxiously waiting for the test results. Carl doesn't even know who she is.

Dr. Tongco came in. He had the test results. Everything looked good. No broken bones. **The blood tests have ruled out any drugs in his system.**

"He is not on any drugs."

Diagnosis: Amnesic shock.

Dr. Tongco decided that it would be best to keep Carl in the hospital for a few days for observation. Carl is taken upstairs to a room. He still does not know who he is. *Who is this woman?*

He still does not know his wife of over seventeen years. Carl calls her, "The pretty lady from the emergency room."

Carl keeps talking of the three-day pills... just floating.

He is having a hard time remembering...

Margery gave Carl a pencil and some paper. She told him to write down anything that he can remember.

Carl drew a picture of the **Being**, on the piece of paper. He then wrote **Ender** and then drew a fork in the road. He then wrote *Saltillo Street*. Carl said, "That, is all he can remember." He was tired.

Margery gets ready to leave; to let Carl get some rest. She gathered up all his clothing to take home.

Roy has stayed all this time and offered to take Margery home. When they arrive at her house, **the sun is just coming up in the East.** Margery makes a comment that she thought the sun had already risen. Both comment that this is very unusual; but, both are tired and too exhausted to give it much more thought.

Roy helped Margery out of the car saying, "If there is anything I can do to help, just let me know. I will be staying in touch."

Margery goes into the house. She was tired and exhausted. She went to put Carl's belongings away, and just go to bed.

Something fell out of the canteen pouch. What is it...? **It... Looks like a piece of mangled metal**.

She put the *mangled piece of metal* on the bureau and went to bed.

The phone rang several times – People wanting to know how Carl is. It is all such a blur... she doesn't remember who, but she told someone about the *piece of mangled metal* she found in Carl's canteen pouch.

They tell her to take it to the sheriff.

☙ ☙ ☙

CHAPTER ELEVEN

After showering and a change of clothes, Margery took *the piece of metal* to the sheriff's office.

The sheriff was out. Margery had to speak with the deputy. She asked him if he knew what *it* was.

The deputy took it from her — looked at it and took it to the back room. He came back saying, "It is a shell casing. I've never seen one in this condition. It looks like it has been turned inside out."

He then took it into another room. In a few minutes, he came back with *the piece of metal*. He handed it back to Margery, and verified, "**It is a 7MM Mag shell**."

Margery asked, "How did it get in this shape?"

The deputy replied, "Just leave it alone. *It was a weird night.*"

"Just — leave — it — alone!"

Margery told the deputy what Carl had written on the paper. She then said, "Enders, I don't know what that means."

The deputy said, "There was an Enders in the forest the same night. But, he was in another area of the forest."

"Don't question it…"

"Just — leave — it — alone!"

❦ ❦ ❦

CHAPTER TWELVE

Carl's memory returned when his oldest daughter, Rose, came into his hospital room. It was like he was programmed to get his memory back when he saw her.

His tests have returned. He is okay. Dr. Tongco has released him from the hospital, and he can go home.

༺ ༺ ༺

After he gets home, he gets a call from the town's newspaper.

The Daily Times, a Rawlins newspaper, has learned of Carl's search party and of his rescue.

Sue Taylor, a reporter for the newspaper, called Carl for an interview. She had heard some of the story surrounding the search and wanted to know if the story was true.

Carl says, "I don't know if you want to talk to a crazy man or not. I must be crazy, but, I was taken 163,000-light-miles from here and was brought back." Carl told Sue what he could remember at this point.

The story makes the front page of the Rawlins newspaper, *The Daily Times*.

ꙮꙮꙮ

Chapter Thirteen

Roy Fleming called Carl from Riverton, and asked if Carl would mind talking with Rick Nantkse, a counselor, about his experience? He told Carl, "Rick is very interested in these types of cases."

Carl said, "I don't mind if Rick doesn't mind talking to a crazy man because I must be crazy. This is just too bizarre..."

Rick called Carl, and they talked of the experience. Rick asked Carl if he would mind talking with Dr. Leo Sprinkle, a psychologist, with *The University of Wyoming* in Laramie, Wyoming. "He is very interested in UFO Experiences," said Rick.

Carl told Rick that he would *like to talk* with Dr. Sprinkle. "Maybe, he can help me sort this out."

&&&

Carl's two daughters, Rose and Lily, and his two sons, Mike and Lyle, go back to school. Margery goes back to work. The family is trying to get back to normal.

Carl is very unsure of himself, and at times he doubts his own sanity. He walks everywhere he goes. He will not drive the pickup. He does not go back to work. He is scared he might harm someone. He awakens at night and swears that he must have dreamed this.

His wife tried to reassure his sanity by reminding him of the *search party*. Search parties are not organized for those dreaming. What about the hospital bill? Dreams are not billed. What about what others saw that night? What about the **bullet**? How did it get in that shape?

THIS WAS NOT A DREAM!

"Marilyn and Don were with me when you were found out south."

"I still have the mangled bullet."

"That is proof..."

"Something—happened."

߷ ߷ ߷

Chapter Fourteen

Margery doesn't know why, but she took the **bullet** to a professional photography studio in Rawlins. She had professional photos taken of the bullet alongside a normal 7MM Mag shell. **WHY**?

She still doesn't know why she had professional pictures taken of the bullet. She is normally a frugal person and doesn't like to spend money foolishly.

WHY? Did Margery feel the urge at 4:00 p.m. that she should find her husband? He was elk hunting. She knew if Carl bagged an elk; it would take time to field dress and carry out. It takes time...

What happened to the bullet he fired?

Why was there no noise when he pulled the trigger?

Why was the forest silent when he fired the bullet?

How did he watch the bullet come out of the rifle, go about fifty feet, and then fall to the ground?

Why was everything in slow motion?

Why did he walk over to the bullet, pick it up and put it in his canteen pouch?

WHY? WHY? WHY? So many questions and no answers...

YET...

🙢 🙢 🙢

Chapter Fifteen

Dr. Sprinkle drove from Laramie to Rawlins to meet with Carl. He drove up to Carl's home at 522 E. Spruce. Carl met him at the door. Dr. Sprinkle introduced himself to Carl and thanked him for taking the time to talk with him of his experience. Carl invited him into his house, and they sat at the dining room table.

Margery poured each a cup of coffee.

Dr. Sprinkle explained that he is interested in Carl's case, and there have been other cases like Carl's. He explained that he is a psychologist at the *University of Wyoming*, and has gotten interested in UFO cases. He would like to study Carl; if he wouldn't mind, to see if they can find more information about his experience.

He asked Carl if he would mind being hypnotized.

Carl said that he wanted to find out all he could about what happened. He is willing to be hypnotized. Dr. Sprinkle also asked if he would mind if he taped the interview for further study.

Carl assured Dr. Sprinkle that he would also like to know if there was more information. He would be glad to be hypnotized; to see if *he* can make any sense of this **Bizarre Experience**.

Margery poured each another cup of coffee. They drank the coffee, and just talked about a little of everything.

The doctor then told Carl to relax, as he started the tape recorder and explained the hypnotic procedures.

ぁぁぁ

Chapter Sixteen

Dr. Sprinkle then gave Carl some relaxation suggestions, and Carl went under. He was hypnotized.

Dr. Sprinkle then asked Carl questions about his hunting trip.

Carl answered all his questions until he came to... the lights...

THE BRIGHT LIGHTS!

Carl's eyes started to water profusely, tears streamed down his face.

Dr. Sprinkle gave Carl a relaxation exercise to help him get past **The Bright Lights!**

Everything is then okay, until, **Auzzo One** brings Carl back to earth; he floated down out of the craft, slipped on the side of the hill and hurt his shoulder. **OH... IT HURTS...**

IT HURTS ... IT HURTS... SOOO BAD!

Dr. Sprinkle gave Carl another relaxation suggestion.

This one helped, but the pain stayed.

Dr. Sprinkle decided at this point that Carl has had enough, and brought Carl back from the hypnosis. But, first, he gave Carl a suggestion that he will be able to remember all the hypnotic adventure. He will be able to distance himself from the experience and will be okay with it.

Carl and Dr. Sprinkle have another cup of coffee and discuss the details that have come forward from this session.

Dr. Sprinkle turned the tape recorder off. He explained that he will have a transcript made of the session and have one sent to Carl.

After another round of coffee, Dr. Sprinkle thanked Carl and Margery for their time and for the opportunity to study Carl. He then left the Higdon's... planning to go back to Laramie.

He is intrigued by Carl's *experience*. Before Dr. Sprinkle left town, he stopped and talked with the sheriff and members of the search party. He talked with the Doctor and the Nurse at the hospital. He talked with other people in town to see what they thought of Carl.

Most of the people in town say that Carl is honest, hardworking, straightforward, matter of fact, type of person. If it had been anyone else, they wouldn't have believed the story.

A few that didn't know Carl told Dr. Sprinkle, "Oh, isn't that the guy they took to the **Lunie Bin**?"

&.&.&.

I know, it is hard to believe. But, a lot of the people in town had also heard of other **weird happenings**, in that area of the forest.

&.&.&.

CHAPTER SEVENTEEN

Carl still doubted himself. He couldn't believe that he could watch a bullet come out of a rifle, and fall to the ground. There must be something wrong with his NEW rifle...

He wants to take the rifle out again, and fire it; to make sure it will penetrate an object.

He and his wife decide to take the snowmobiles, make a family outing, and go **North of town.** Carl does not want to go back into the area that *IT* happened.

He loaded up his **NEW** 7MM MAG rifle, and they headed north of town. When they got to the forest, Carl took his rifle from its case, pointed, and fired it at a tree... **IT PENETRATED THE TREE...** *like it should*.

There was nothing wrong with his rifle. Everything is normal. Carl felt relieved. They all relaxed and had a good outing, riding their snowmobiles, playing in the snow, and walking amongst the trees.

Watching a deer every now and then. Enjoying nature as usual.

They now have a normal, *Higdon outing*.

ଈଈଈ

CHAPTER EIGHTEEN

Dr. Sprinkle was intrigued with Carl's **experience** and would like to study him further.

He called Carl for another appointment. He would like to put Carl under hypnosis again to see if more information would come forward.

Carl told Dr. Sprinkle that he is remembering more and more of the events that happened.

Dr. Sprinkle asked Carl if he would mind if Rick Kenyon, a Rawlins Art Teacher, came to the house and drew some sketches of the **Being**?

Carl agreed that it would be nice to see an artist's conception of the **Being**.

Dr. Sprinkle said he would get back in touch with Rick to set up an appointment.

❦ ❦ ❦

Chapter Nineteen

Dr. Sprinkle again drove to the Higdon house in Rawlins to meet with Carl for further information. He met Rick, waiting in his car, in front of the Higdon's house.

He and Rick walked up to the Higdon house together.

He introduced Rick to Carl, and Carl welcomed them into his home. The three sat at the dining room table.

Dr. Sprinkle discussed the case with Rick, as they each had a cup of coffee. Rick is intrigued by Carl's story and asked him if he would mind if he sat in on the session. Carl welcomed Rick to sit in and told him maybe he could shed some light on what had happened.

Dr. Sprinkle again gave Carl relaxation suggestions. Carl went under. Again, Dr. Sprinkle asked Carl of his trip with **The Aliens**. Carl remembers more and more. He remembered that there were other people standing under the tower of **Bright Lights**. He remembered that he and two Aliens **float** out of the craft and into the building. He remembered that he was placed in a room and a **wall** came down in front of him and was then taken away. **The Beings** then told him, he was not what they wanted, and they would take him back. When Carl was brought out of the building, he noticed some people standing under the tower. They seemed to be human, like Carl. They seemed to be at ease as if they had been there for some time.

Carl and only one **Being,** this time, **floated** back to the craft. He told Carl his name was **Auzzo One**.

There is a dark void. Then they are over the forest. Carl is floating down out of the craft onto a side of a hill.

He slips and falls.

IT... HURTS... MY... SHOULDER HURTS ... SO BAD!

Dr. Sprinkle gave Carl a suggestion that he could get past the pain. The pain eased, but it would not go away. He then brought Carl out of the hypnotic state. He gave him a suggestion that the pain would ease more and more, and he would be able to remember everything and would be okay with it. Dr. Sprinkle then turned off the tape recorder.

Margery poured each of the men another cup of coffee. They discussed the new developments that came from this session, and how the **being** looked, as they drank their coffee.

Rick proceeded to sketch as Carl described the **Being**.

"He was six feet tall, had no chin, had bowed legs, straw-colored hair with two strands sticking straight up, like antennas. He had no visible ears. He had slanted eyes. His skin coloring is the color of an oriental. **The Being** had two arms, but, no hands. The one arm came down... and then nothing. The other arm had a cone-like device on the end of it. Whenever he pointed and waved the cone-shaped device, the item disappears. I don't know where it goes. You just can't see it anymore. He is dressed in black. It looks like what a scuba diver would wear."

When rick finished the sketch of the **Being**, Carl remarked, *'That looks just like him.'*

Dr. Sprinkle made another appointment for more hypnoses and studies of Carl.

He arranged for Carl to undergo various tests at the *University of Wyoming.*

🙦 🙦 🙦

Psychological tests are given. Carl undergoes Physiatrist testing.

Test results:

HE HAS NO MENTAL ILLNESS.

APRO also does a study of Carl.

Dr. Hyneck tried to debunk Carl's story, but in doing so, only gave it more validity.

Carl underwent more and more testing. He wants to find out what happened, as much as they do.

Many times, Carl called Dr. Sprinkle. He would help Carl with relaxation exercises to get past the pain, and sometimes to just talk.

It is still hard for Carl to believe what happened to him.

&&&

Dr. Sprinkle came to the Higdon house many more times; sometimes Carl and Margery drove to Laramie to Dr.

Sprinkle's house to talk, or to the University for further testing.

This went on for months and months.

🙦🙦🙦

Chapter Twenty

Margery was in the kitchen doing dishes. Carl was watching TV in the living room. He called to her and suggested they go for a drive.

Margery is anxious to do anything to get out of doing the dishes, so she yelled back, "Sounds good to me." She wiped her hands and got her jacket. Their son, Mike, was home and they asked him to go with them.

Carl drove south of town. He seemed to know exactly where he was going. He drove several miles south, took a dirt road to the west, and up a hill.

As Carl, Margery, and Mike arrived, there was a bright **green light** in the sky. It looked like a large upside-down ice cream cone.

Then the smells... terrible, horrible smells... like dirty, dirty socks and then Sulphur. The smells were **HORRIBLE!**

Carl drove to the top of the hill and shut the motor off. He said, "I am late. I was to be here a few minutes earlier."

We sit there for a few minutes... when Carl hit the top of the steering wheel with his fists. "I must be crazy. Why am I here? If we had been here on time what would have happened to us?" Carl started the car and back to town we went.

THANK GOD, WE WILL NEVER KNOW WHAT WAS MEANT TO HAPPEN!

Several nights later, Carl woke his wife; and asked her to put her hand on him. He said, "I know it is weird. I am in bed with you; yet, I am out south of town. They are telling me, "Look for a Black Box." They are taking wild game. It is like they want me to know that they can contact me whenever they want. "**WHAT IS WRONG WITH ME?**"

Carl is so puzzled with this new development that he called Dr. Sprinkle; even though it is the middle of the night. He is sure he is going out of his mind. He told Dr. Sprinkle of the experience and how his shoulder **hurts — bad!**

Dr. Sprinkle gave Carl a relaxation suggestion, which eased the pain... some. Dr. Sprinkle asked to set up another hypnotic session with Carl.

Over the years, Carl will be calling Dr. Sprinkle many times. The events that happen are so puzzling.

Why do they want to let Carl know that they can contact him anytime they want? The pain in his shoulder seems to get worse each time they contact him.

On numerous occasions, on his way to and from work, there have been lights following his pickup. This has also been seen by his crew. The crew has finally gotten used to it. They don't even pay any attention to them anymore.

The science teacher at the high school called. He asked if his class could come out to the house and meet with Carl.

The town people are more aware of the strange lights in the skies.

Carl gets calls from all over the world, Canada, England, Australia, etc., and of course various point of the U.S.A. Most of the callers just want to know if there really is a Carl Higdon living in Rawlins, Wyoming; since there is, they get the phone number and talk with him.

Carl talks with the callers. He figures it is their dime, and since they are calling from such a great distance they must be interested in the **experience,** and **he needs to talk about it**.

Many that call have also seen **strange lights**. They tell Carl, "It is so good to finally be able to talk with someone about it and have someone to understand, and **not** tell you, "*You are crazy*."

Carl tells them that he can relate to that. At times, he doubts his own sanity.

Carl then has more confidence in himself, and he went back to work.

The bullet still has Carl confused. How did it get so mangled? It looks like it has been turned inside out... where is the lead?

Carl gets to wondering about the lead. He decided that he will get a metal detector and go back into the area that it happened; to see if he can find the lead with the metal detector. He decided that maybe now is the time that he should. Their friend Don and Marilyn state that they would like to go with him into the area to help look for the lead. They would like to see what else they can find. Some other friends say they would also like to go with them.

They load up their families and some food in several pickups. They decided to make an outing of it, and head south to the forest.

In the forest, they park their pickups on the hill and walk down the same winding road that Carl had taken; taking pictures of the kids as they stood in the **deep ruts** that the search party had left; when they rescued Carl.

They take pictures of the **muck hole,** where Carl's pickup had been.

They take pictures of where the **cubicle** had set, and of the foliage around, and how the leaves were burnt.

They take pictures of the **strange carvings** on the trees. The carvings looked like they had been there for a long time. But, what do they mean?

Carl ran a metal detector over the area, but, no lead could be found...

&. &. &.

CHAPTER TWENTY-ONE

In Search Of, TV movie crew came to Rawlins and did a movie segment of the *incident*. They went out close to the area, *but*, the camera crew *refused to go into the actual area*. The crew said, *"It is too eerie."*

The National Enquirer received news of the story, and they came to Rawlins. Their crew took pictures of *the actual area*. They asked Carl if he would fly out to California to take a lie detector test and to run some other tests; which he did.

A Japanese film crew came from Japan and did news clips on Carl and his *experience*. The Japanese film crew, also, would not go into the actual area. They said the area, did not feel **right**.

Carl's story has been in many national papers and magazines...

In Japan, we have been told, they have erected a Space Museum. We are told that in the museum, you have only to push a button, and you can hear the story of Carl's trip into space.

His story is in *Time-Life*. It is on database software and is in the *UFO Encyclopedia*.

You can go to the search engines (*Google, Bing, Yahoo*, etc.) on the internet, type in Carl Higdon and get his story. The story of the *incident* is still on the internet, **after forty-three years**. Some of the stories are factual, and some are dramatized.

His story is mentioned in movies and books that we are unaware of; unless someone sends a copy or calls us. Several times we have received calls to see if what *someone* has read or seen about Carl is true.

We have not seen all the articles or movies.

WE CANNOT SAY IF WHAT THEY SAW OR HEARD IS TRUE.

We can only say, "YES — IT DID HAPPEN!"

ߜ ߜ ߜ

AFTER THE EXPERIENCE

We have been asked by many: Did you and your family make money from this experience?

No – *We did not make money* from this experience. We *lost* money. It cost us *time, money, and nervous emotions*; for Carl went through various, exhausting testing, and numerous lie detector testing. Carl and his wife, Margery, both, took time off from their work for Carl's tests, **without** pay.

Margery went with Carl each time he had testing. She wanted to check what kind of tests they were doing with her husband and stayed with him during each of his tests, and interviews.

Carl went through various testing and interviews by different UFO investigators. He also took written tests; for **he – himself –** *also wanted to know more answers about this experience.*

He thought there *must be something wrong with him*. The experience was *hard for him to believe*. At times, he doubted his own sanity.

BUT, *there is the proof:*

The search party that found Carl in the forest, dazed and in an amnesic shock condition.

The muck hole, Carl's two-wheel drive pickup was in. *No* tracks led to where the pickup was stuck. (How did it get there?)

The Game Warden that talked with Carl earlier that day on top of the hill — *with his pickup.*

The *lights* that were seen by others that night.

Doctor Tongo's medical report when the search party brought Carl to the hospital.

AND THEN THERE IS THE MANGLED BULLET!

Carl shot his 7mm mag rifle, at the elk, watched the bullet in mid-air. It then hit something and fall to the ground. He walked over, picked it up and put in his canteen pocket. The fact that Margery had professional pictures taken of the bullet – One with the shell, and one with a regular 7mm mag bullet. Dr. Walter Walker, APRO's Consultant in Metallurgy, examined the mangled bullet, and could only say that it hit something with great force to turn it inside out. No lead was found.

Dr. Angela L. Howdeshell M.D. of the *University of Wyoming* did a psychiatric evaluation to determine Carl's emotional and mental health.

RESULTS:

NO PSYCHIATRIC ILLNESS

ɞɞɞ

Robert J Hudek, Ph.D. biological scientist, tried to debunk Carl's story but it seemed the more he tried to debunk the story, the more credibility he gave it...

The Being having no eyebrows — is related to an environment that is low in ultraviolet light.

The Being wearing a black suit to shield him from our sun, because it burned them...

A vitamin D deficiency, devoid of sun, would give rickets — bowed legs and skeleton

❦ ❦ ❦

Carl walked everywhere he went, after the experience. He would not drive the pickup. He was unsure of himself. He was scared he might hurt someone.

Talking with Dr. Sprinkle, his hypnosis, and all the testing and interviews, Carl went through, helped him to get his confidence back, and to go back to work, and on with his life.

In Search Of Film crew, came to Rawlins, Wyoming, and filmed a segment on Carl's experience. I like the way Carl ended the film:

"If you have an experience; talk with someone you can trust."

"Don't dwell on it."

"Then get on with your life."

This is the best suggestion we can give!

Carl had been passing numerous kidney stones before the experience. He had scars on his lungs. Since the experience, he no longer has kidney stones, and the scars are gone from his lungs.

The photography studio that took the pictures of the bullet burned shortly thereafter.

The **bullet** disappeared from a safe at the *University of Wyoming* – but not until it had been examined by a metallurgist.

Our telephone seemed to be tapped.

These kinds of things seem to happen with **UFO experiences**.

But, before the bullet disappeared Margery had sent the studio pictures of the bullet to different places throughout the world. Again, why did she do this?

She does not know. Still so many questions and still so few answers.

Maybe, some-day — we will know?

MAYBE — MAYBE SOMEDAY...

WE WILL KNOW... WHY

UNTIL THEN,

CARL SUGGESTS, IF THIS HAPPENS TO YOU...

TALK WITH SOMEONE YOU CAN TRUST,

DO NOT DWELL ON IT,

AND GET ON WITH YOUR LIFE.

☙☙☙

DO I BELIEVE IT HAPPENED?

I WAS THERE...

I was in the forest when the search party found him. I saw the condition he was in, cold, in shock and confused. I saw the strange lights in the forest. The forest was so light you could pick up a dime from the road, but there was no full moon. I saw the lights in the East that looked like the sun was coming up, way too early for a sunrise. I watched the actual sunrise hours later when it was the appropriate time.

I was there when the search party brought him from the forest to the ER at the hospital in Rawlins, Wyoming.

I was there when the nurse examined him in the ER. I saw the tears in his eyes, water streaming down his face, from the bright lights. (The ER lights and the previous lights, **from the other place**.) I saw and heard him with his shoulder pain.

I was there when the ER doctor examined him and took tests for drugs in his system. The doctor had said, "Surely he must be on drugs, for the things he is talking about."

This new doctor to the Rawlins area took notes of what Carl was saying: Of the **Men in black** taking him **163,000 light miles**. Of watching his pickup **disappear** from the hill where he parked it. Of watching a **blue marble** shaped ball below him. How he didn't walk; he only **floated**. How they took his **elk**. This ER doctor is so enthused by what Carl is saying he makes notes of it in his medical report.

I was there when Carl did not recognize me, his wife of over sixteen years, the mother of our four kids.

I was there when he asked, "Who is Carl? They keep calling me Carl, but, I don't know who he is."

I sat with him in the ER and again when he was taken to a room upstairs.

I was there when he came home from the hospital, so unsure of himself. He walked everywhere he went. He was scared to drive the pickup. He was afraid to go back to work, worried he might get someone hurt.

I was there when Dr. Sprinkle visited him and hypnotized him numerous times, about that incident.

I was there when Mr. Kenyon drew sketches of Auzzo One, as Carl described the being.

I was there when he was given test after test, and also, the lie detector tests.

I was there when he was asked question after question about the incident, as **UFO Researchers** questioned him.

I was there when he questioned himself.

I was there when we went back to **the place** and took pictures of the area where it happened. I saw the ruts the search crew made when they went in to get Carl with four-wheel drives – Carl's pickup was a two-wheel drive. The crew had to put logs to get Carl's pickup out of mud hole where his pickup had been deposited – where only a four-wheel drive could go, and then with difficulty. I saw where he shot at the elk - Where the vegetation had been seared with high heat.

I was there when he was found and saw the condition he was in. Carl is a very strong-willed, matter of fact, man. Not much bothers him. I was in the hospital with him, I was with him for all his testing, and then getting on with his life over all these years.

If you want to believe, it or not, he feels that is your choice.

He knows, **IT HAPPENED TO HIM!**

YES! I believe!

This year Carl and I have been married 59 years.

His wife,

Margery Higdon

☙ ☙ ☙

Margery's Account and Emotions of That Experience...

I wrote Carl's story over thirty years ago. I wrote it as therapy for **me**. I would get this strange, uneasy feeling every time someone would bring up his story or talk about it. Like the **inside of me is shaking**. I felt that I **had** to write the story, to try to understand what had happened to my husband; and to help **me** to understand why I did some of the things I did.

To this day, some forty-three years later, I still get this shaking feeling inside, when I talk about his story.

I have been asked by many; how did you feel? What were your emotions?

ઢ ઢ ઢ

When I came home from work, the afternoon of October 25, 1974, at 4:00 p.m., I could **feel** that Carl was in trouble and **needed** help. (We have this close connection.) I had this **obsessive urge** that I had to go to him. I could feel the **uneasy feeling** in the pit of my stomach. The top part of my chest felt like **someone was squeezing me**. It became an **obsession**. I **had** to go to **him**.

Later I found this was around the time he took his trip, *"Out of this world."*

&&&

After I talked with my girlfriend, around 6:00 p.m., the uneasiness eased. I busied myself making supper for the family, and the uneasiness subsided. I felt then he would be **okay**.

By the time Bud called around 6:30 p.m., I was fairly **calm**.

I still felt the feeling that I needed to go to him, but I was not as strong now. Now, I just **needed to see him**, to *know* he was okay.

The waiting in the forest seemed to take forever. When you are waiting for your loved one, I think it does take forever. I was **anxious,** and yet, **apprehensive;** not knowing if Carl was hurt or what condition the party would find him in, Waiting---and needing to know.

Happy and anxious to see him when he is found. Then, **concerned and confused**, for the condition he is in.

The waiting at the hospital was **agonizing.** There was nothing I could do to help him, but sit, and wait beside him. I was **worried.** What happened to make him this condition? He didn't seem to be physically hurt. Why doesn't he know me? I 'm his wife of over sixteen years. I am the mother of our four kids. What is wrong? Why can't he remember?

So **happy** and **relieved** when the Doctor came back with the medical reports; no broken bones, no drugs in his system.

I was **worried** and yet **curious;** why can't he remember. What happened? Will he be able to tell me, to let me know?

I was **concerned** when he regained his memory—and knew his family, but insisted he must be crazy or insane.

Amnesic shock — what happened? He is physically strong, very strong. He is a matter of fact, curious, and strong-willed man. What could put him in this condition?

I was **scared**, but **curious**, to know — what put him in this condition.

He'd gone hunting, got lost and was found, what more could it be? He must be hurt worse than his body appears. But — WHAT?

I decided to give him a pencil and some paper, maybe he could let us know.

But, what he wrote – and drew – what did it mean? Would this let me know what happened to him... so confusing?

He is tired, and I'm **exhausted.** I convince myself he is safe in the hospital, and I just want to go home and get some rest.

The bullet has me **puzzled.** It looks like a piece of mangled metal. Why would Carl put it in his canteen pouch? What does it mean? Does it mean — anything? I'm too tired and exhausted to think any more of it. I put it on the bureau and go to bed. I'm so...**tired and exhausted.**

After several calls about his condition, I tell someone of the strange metal...I'm told to take it to the sheriff. I pick up the metal, turning it over and over. I'm **curious.** What is this? Would the sheriff even know? Does it have anything to do with Carl's condition?

After taking the shell to the sheriff's office, I'm more **confused**. It is a **7MM Mag shell.** The kind that Carl's rifle shoots. How can this be? What happened in the forest? Now I'm **confused** and **curious**. I **need** to find out more.

But, the deputy tells me**, "Leave it alone, just leave it alone. It was a weird night!"**

Come on, that just **gets my curiosity up more;** for at this time I really had no idea of what had happened.

Carl keeps talking of the men in black, "They took my elk!" What does this have to do with a bullet turned inside out? I'm **worried** about Carl's condition and let my thoughts of the bullet slide to another day.

When Carl came home from the hospital, he kept repeating, "I must be crazy, this is too bizarre!" *He, himself, can't believe* what happened. He is positive he had to have dreamed it. The memories that are coming back to him are just too bizarre.

I'm **confused and worried** about Carl. I keep reassuring him of the details of the search party, the hospital stay, and **THE BULLETT HE SHOT.**

The story of the search party, his rescue, and his interview have made the daily newspaper. People in town are talking of the rescue and commenting on other weird happenings, in that area of the forest. Most of the people believe Carl. They know him to be hardworking, honest, and a matter-of-fact person.

I still had the kids to worry about and **had to stay calm.** The next several days are a blur as we try to get on with our lives.

The art teacher asked to come to the house to make a drawing of the **BEING,** from Carl's description. While he was there, he talked of other weird experiences in that area of the forest. He talked of Dr. Sprinkle, a Phycologist from the *University of Wyoming*, who studied 'UFO's and strange, weird experiences. He asked if Carl would be willing to talk with him.

I am so **thankful** that Rick came to the house; of the drawing that he drew, and for suggesting Carl talk with Dr. Sprinkle.

I don't think we could have made it without them; listening to Carl and letting him talk of his experience. The testing he went through with different organizations. Finding out that it has happened to others. Carl finding out he did not dream it, (although I think he would rather have dreamed it, then to have gone through it.) **Relief** – he was not crazy.

Then, there was **THE BULLET**, proof something happened.

&.&.&.

We have gone on with our lives. Our kids have all graduated from school and now have families of their own.

Carl went back to work in the oilfield – retiring in 1997, after forty-eight years in the field.

I went back to work for thirty-two years, working as secretary-bookkeeper, and eventually as an administrative assistant. During this time, I even went to college and received my diploma in business. In 1997 I was forced to retire due to major surgery.

In 2002, Carl and I built our all-steel house, just he and I, no contractors.

We now live in our home we built and enjoying life.

❦❦❦

REMEMBER
IF THIS HAPPENS TO YOU
TALK WITH SOMEONE YOU CAN TRUST

DON'T DWELL ON IT
AND
GET ON WITH YOUR LIFE

❦❦❦

CARL

Carl is now in his eighties. He has macular-degeneration of his eyes and has a heart condition.

He is retired from the oilfield, and enjoying life.

He and his wife have four kids and have been married for fifty-nine years.

Carl and Margery lost their oldest daughter, Rose with cancer this year.

Carl and Margery are grandparents and great-grandparents.

&.&.&.

Made in United States
Orlando, FL
27 May 2023

33535539R00046